VINTAGE MATCHBOXES

A LITTLE STICKER BOOK

Skittledog

SAFETY MATCHES
MADE IN
KOBE JAPAN
壽
怡和洋行

火
虎
柴
標

SUPERIOR SAFETYMATCHES
TRADE
MARK
MADE IN TAKASHIMA. JAPAN

LA FLÈCHE ET LE TIGRE
MADE IN
SWEDEN
ALLUMETTES DE SÛRETÉ

BEST MATCHES
MADE IN JAPAN

PROTECTION
FROM
FIRE
TAKASHIMA
MADE IN JAPAN

SUPERIOR SAFETY MATCHES
RIOSUI & CO.
MADE IN JAPAN

SAFETY MATCHES
利輿公司
MADE IN JAPAN

SAFETY MATCHES MADE IN JAPAN

蓮猴商標
宏利廠造

BEST MATCHES
MADE IN JAPAN

BEST SAFETY MATCHES
MADE IN JAPAN
猩々酒爲記

Mabe in
Japan

METRO
PRICE 0-0-6
ROBIN BRAND
SAFETY MATCHES
METRO MATCH Co.
POONITHURA

PENGUIN
T. M. Regd
308433
50's MATCHES PRICE 13 Ps.
SRI BALAMURUGAN MATCH WORKS
PERNAICKENPATTI. (VIA) SIVAKASI.

LOVE BIRD
VASANTHA MATCH WORKS
RAJAPALAYAM
SAFETY MATCHES

TWO PEACOCKS
PRICE RE. 0.04
THE PALANIAPPA MATCH
INDUSTRIES, SIVAKASI.

RAINPROOF
SAFETY MATCHES
MADE IN SWEDEN

LUCKY BIRD
RADHA MATCH WORKS SATTUR
SOUTH INDIA

THE
CONDOR
SAFETY MATCH
MADE IN SWEDEN

OWL BRAND
TRADE
MARK
SAFETY MATCHES
MADE IN JAPAN

Safety Matches
FUNAI MATCH WORKS

BLUE BIRD
नीला चिड़िया
PRICE 0-03 NP
SYED MATCH WORKS
VIRUDHUNAGAR.

LOVE BIRDS
ANDAVAR MATCHWORKS
SADHANANDARURAM

THE INSEPARABLES
MADE IN JAPAN
SAMUEL SAMUEL & CO. LTD. KOBE

BLUE BIRD
PRICE Re 0-06.
PERINTALMANNA RICE & OIL MILLS. PERINTALMANNA
SAFETY MATCHES

NIGHTINGALE
NIMCO
PRICE Re. 0-0-6
NATIONAL INDIAN MATCH CO. DINDIGUL
SAFETY MATCHES

BULBUL
PRICE
RE.0-06
VEERABADRA MATCH FACTORY,
SATTUR

SAFETY

HUNTING TIGER
PRICE
SIX PIES
SAFETY MATCHES
SUGAR MATCH WORKS SATTUR
JAI HIND MATCH DISTRIBUTING COMPANY SURAT.

DOUBLE COWS
SAFETY MATCHES
MURALI MATCH WORKS
CHITTOOR.(ANDHRA)

SAFETY MATCHES
BEST QUALITY
TRADE MARK
MADE IN JAPAN

INDIAN ANTELOPE
SAFETY
MATCHES
MADE AT TIDAHOLM, SWEDEN

IMPREGNATED
CUPPLES
RHINO
SAFETY MATCHES
MADE IN SWEDEN

CHITTOOR CHEETA
SAFETY MATCHES
PRICE 0.06 NP
RAYALSEEMA MATCH WORKS
CHITTOOR [A.P.]

SPIDER
PRICE Re. 0·0·6
NAVA INDIA MATCH WORKS
THAYILPATTI.
SAFETY MATCHES

THE MONKEY
SAFETY MATCHES
MADE IN SWEDEN

SAFETY MATCHES
NAM SANG WOO.
SAMARANG, JAVA.

豹駝
MADE IN JAPAN

50's SAFETY MATCHES PRICE 15 Ps.
SEA KING
RAMESH MATCHES, PERIAGOLLAPATTI

CROCODILE & SERPENT
TRADE MARK
MADE IN JAPAN
SAFETY
MATCHES
IMPREGNATED
SOLE AGENTS:
HAMIR HASSUM, ZANZIBAR

神戸松谷燐寸製造所製造

四兎扇記
MADE IN JAPAN

ALMACEN
MATCHES

GATO
IN SWEDEN

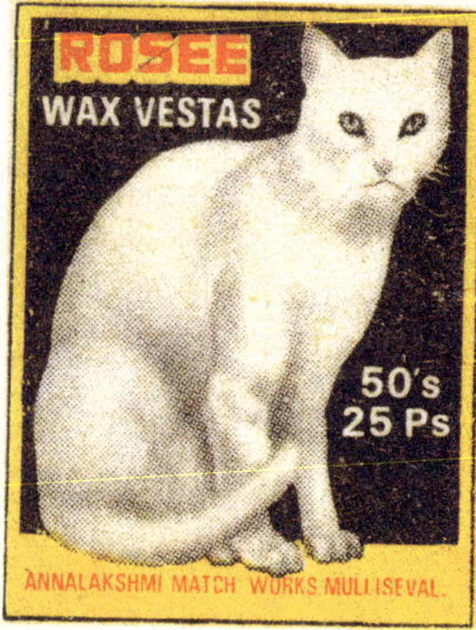
ROSEE
WAX VESTAS
50's
25 Ps
ANNALAKSHMI MATCH WORKS MULLISEVAL.

BILLY BRAND
PRICE 0.06
THE STANDARD MATCH Co. FEROKE
SAFETY MATCHES

JOLLY CAT
PRICE RE. 0-06.
RAJAM MATCH INDUSTRIES
SATTUR.

SAFETY
MATCHES

CAT
KMF
SAFETY MATCHES

CAT BRAND
PRICE
Re.0-0-6
JAYAM
JAYAM MATCH WORKS, SIVAKASI

SAFETY MATCHES
MADE IN JAPAN

大阪近江岸製造

BLACK CAT
IMPREGNATED
IMPREGNATED
SAFETY MATCH

WE THREE
SAFETY MATCHES
50's PRICE 25 Ps
SRI NARAYANA MATCH WORKS
LAKKIDI

TRADE MARK
錦章洋行

SAFETY MATCH
MURAKAMI. OSAKA

SAFETY MATCHES
MADE IN JAPAN

ELAYIRAMPANNAI SM.

SERVICE INDL. CO OP.

SINDC
LL MATCH PRODUCERS

FLYING DRAGON
SAFETY MATCHES
MADE IN JAPAN
M. NAOKI. KOBE
飛龍爲記

SAFETY MATCH
MADE IN JAPAN

SAFETY
MATCH
SEISEISHA:&CO.

TWO PARROTS
PRICE RE 0.03
THE PALANIAPPA MATCH
INDUSTRIES, SIVAKASI.

MADE IN JAPAN

RAJA PARROT
SAFETY MATCHES
50's PRICE 15 Ps.
SARAVANA MATCH FACTORY
ILAYANGUDI.

SAFETY MATCH
MADE IN SWEDEN

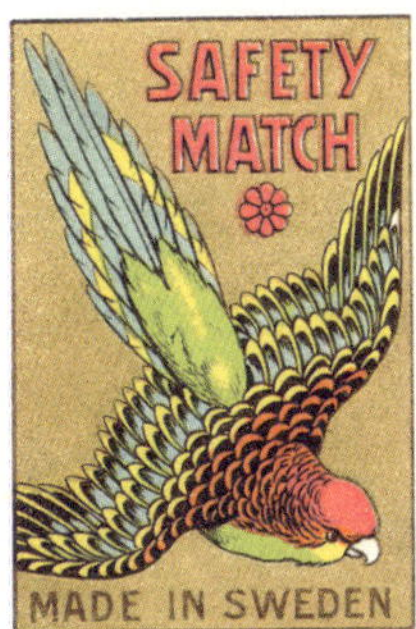
SAFETY
MATCH
MADE IN SWEDEN

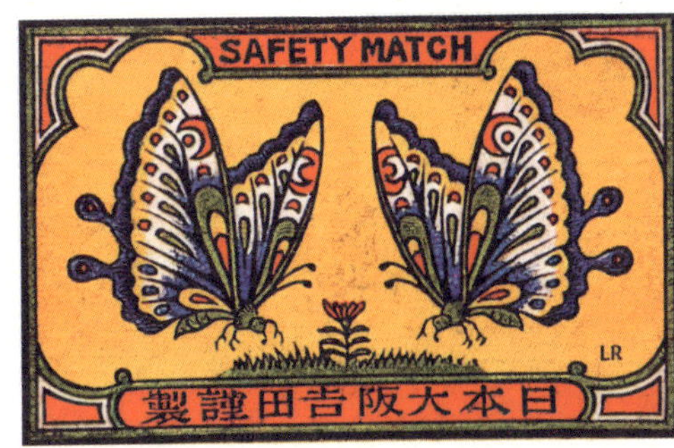
SAFETY MATCH
LR
日本大阪吉田謹製

BUTTER FLY
PRICE RE. 0-05
LAKSHMI MATCH
WOAKS SATTUR.

漢
興
RIOSUI & CO.

MEPHISTOPHELES
SAFETY MATCHES
MANUFACTURED IN SWEDEN.

SAFETY
MATCH
MADE
IN
JAPAN

GIRL & MOON
SAFETY MATCHES
MADE IN SWEDEN

IMPREGNATED
SAFETY MATCHES
MADE IN FINLAND
CARMEN

FLUTE BOY
SAFETY MATCHES
MADE IN SWEDEN

WOMAN & TIGERS
SAFETY MATCHES
MADE IN SWEDEN

MY LADY
MANI MATCH WORKS
DEVAKOTTAI
SAFETY MATCHES

WATER LILLY
A.P.R.S.P.
NATIONAL MATCH WORKS SIVAKASI
SAFETY MATCHES

ENGLISH LADY
SAFETY MATCHES
MADE IN SWEDEN

MAHATMA GANDHI
MADE IN JAPAN

LOVELY GIRL
SAFETY MATCHES
MADE IN SWEDEN

THE
HAMMOCK
SAFETY MATCH
MADE IN
SWEDEN

轆輪安全火柴
MADE IN
EUROPE
商
標

歐洲製造
註冊

TRADE MARK
HIOGO JAPAN

神戸勉榮組謹製

中國
昌明廠造

SAFETY MATCH
優良細軸

SAFETY MATCHES
MADE IN JAPAN
DAWOODALLY & CO.

三鼠爲記
上海益盛廠製

SAFETY MATCHES
MADE IN JAPAN

THE FINEST QUALITY
PREPARED BY N.KITA.
MADE IN JAPAN

SAFFTY MATCH
姫路市高嶋監製

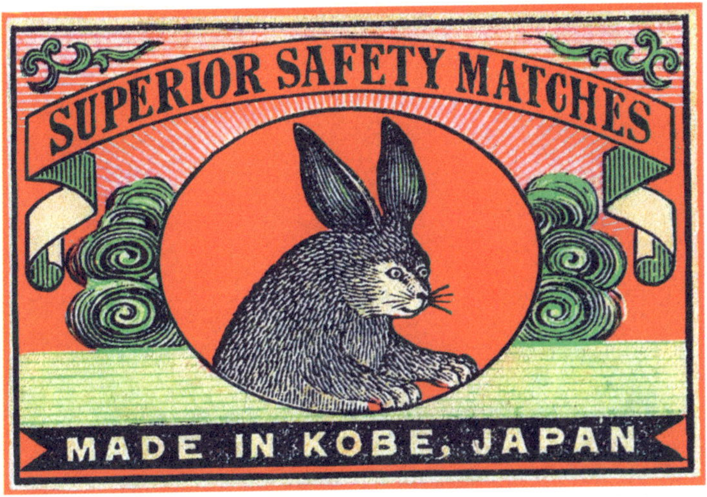
SUPERIOR SAFETY MATCHES
MADE IN KOBE, JAPAN

MADE IN KOBE JAPAN

改良火柴
裕貞祥行監製

TRADE
MARK
NISSHINSHA. MADE IN JAPAN

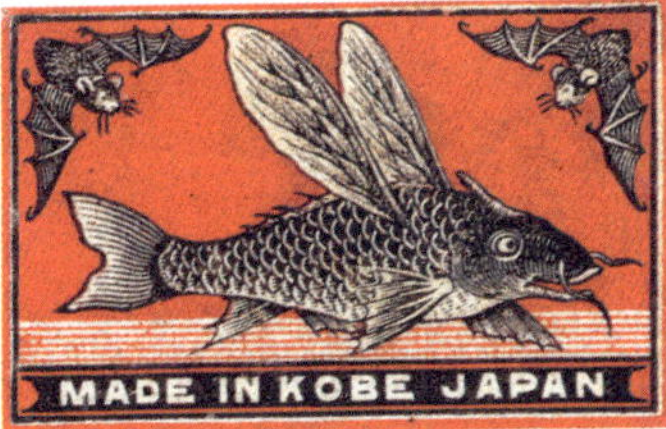

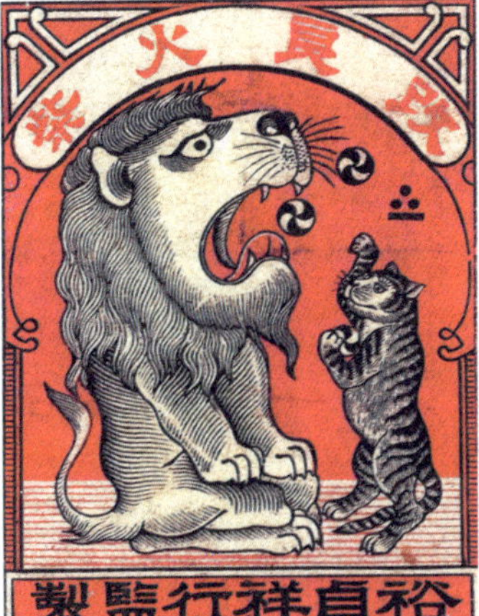

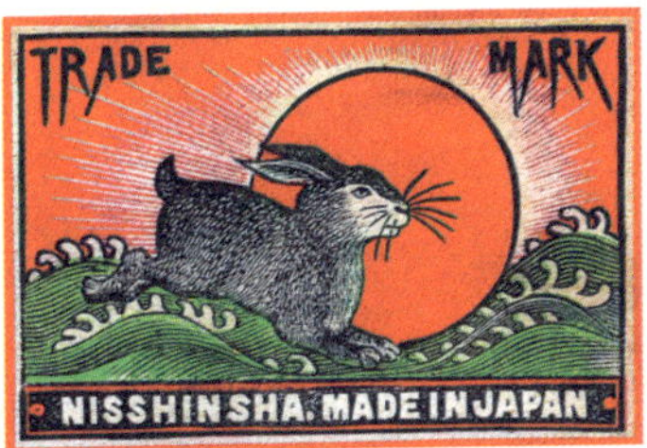

D.M.
Co.
LTD.
MONUMENT

WATER FALLS
THENADAR MATCH WORKS SIVAKASI
SAFETY MATCHES

GARDEN
NEW BHARATH MATCH WORKS, VILATHIKULAM.
SAFETY MATCHES

THE
SWEDISH FLOWER
SAFETY MATCHES
MADE IN SWEDEN

CAPE JASMINE
SAFETY MATCH
IMPREGNATED

CHAMPA FLOWER REGD. No
PRICE RE.0-0-6 142842.
ராம்நாட் மேச் இண்டஸ்ட்ரீஸ்
தம்மநாயக்கன்பட்டி.
MOHANDASGOPALDAS
SIVAKASI.

SUNFLOWER
SAFETY MATCH
SPECIALLY MADE FOR
R.H. GANNI KARACHI
MADE IN YUGOSLAVIA

COTTONFLOWER
60s PRICE RE.0-06
NEWINDIAMATCHWORKS
SATTUR.

फूल
UMCO
PRICE
RE.0-04
UNITED MATCH COMPANY
KADAMBUR

ROSE BRAND
KERALA MATCH FACTORY
MADE IN INDIA
KMF
SAFETY MATCHES

ROSE
PRICE
REO-04
REGD No.
113227
SAFETY MATCHES
EAST INDIA MATCH FACTORY
KOVILPATTI.S.I.

THE BOUQUET
MADE IN
SWEDEN
SAFETY MATCHES

THREE PALMS BRAND
TRADE MARK
MADE IN ITALY
IMPREGNATED SAFETY MATCHES
AVERAGE CONTENTS 60 MATCHES

THREE FLOWERS
PRICE RE
0-0-6
NAVAMONIMATCHWORKS
SATHIYANAGARAM.

MAGIC
HEAD
PRICE
0-0-6
YESPIMATCHTRICHUR

THE HEART
TRADE-
MARK
MADE IN SWEDEN
SAFETY MATCHES

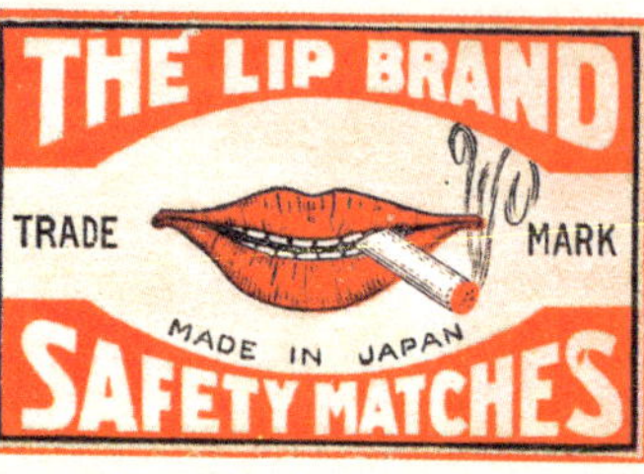
THE LIP BRAND
TRADE
MARK
MADE IN JAPAN
SAFETY MATCHES

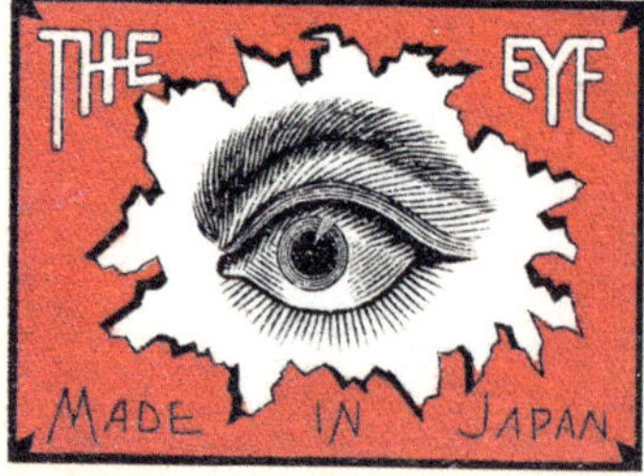
THE
EYE
MADE IN JAPAN

THE HEART
MADE IN
SWEDEN
TANDSTICKOR

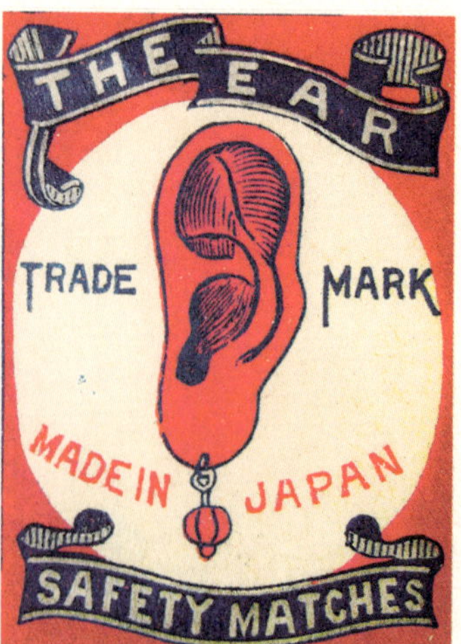
THE EAR
TRADE
MARK
MADE IN
JAPAN
SAFETY MATCHES

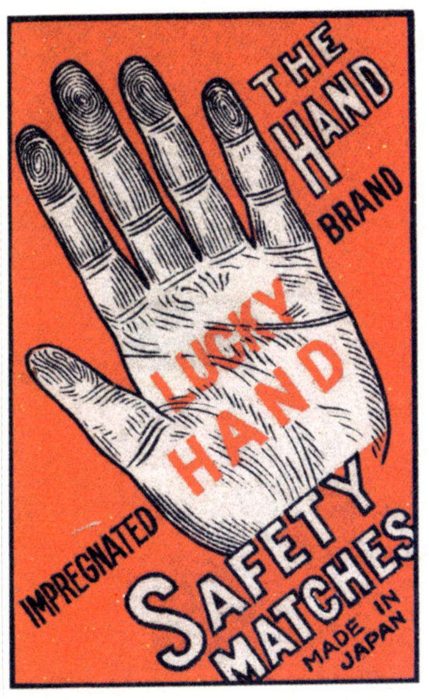
THE
HAND
BRAND
LUCKY
HAND
IMPREGNATED
SAFETY
MATCHES
MADE IN
JAPAN

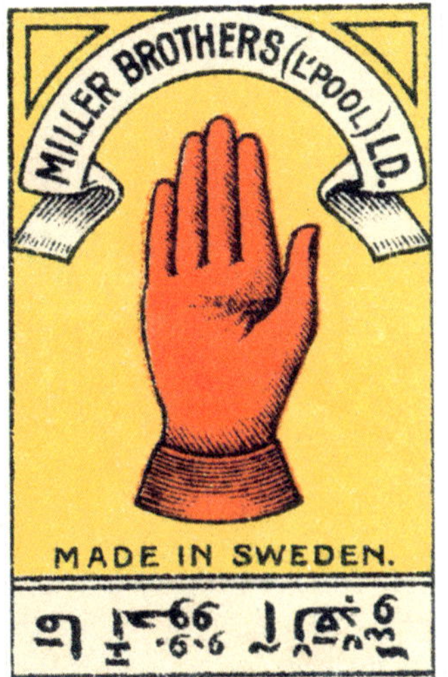
MILLER BROTHERS (L'POOL) LD.
MADE IN SWEDEN.

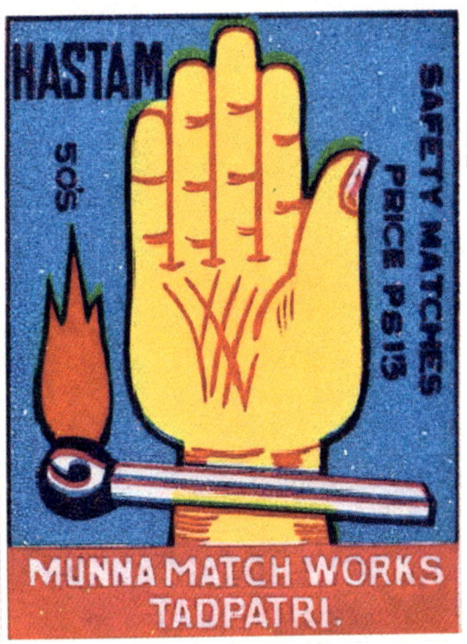
HASTAM
SAFETY MATCHES
PRICE PS 13
MUNNA MATCH WORKS
TADPATRI.

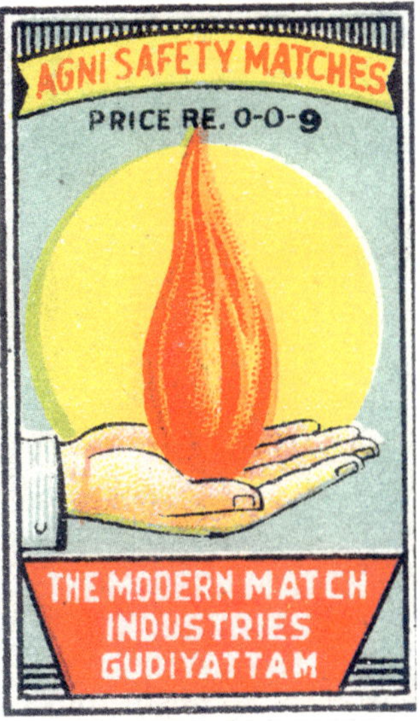
AGNI SAFETY MATCHES
PRICE RE. 0-0-9
THE MODERN MATCH
INDUSTRIES
GUDIYATTAM

REGISTERED
MADE
SPECIAL

TRADE MARK
SWEDEN

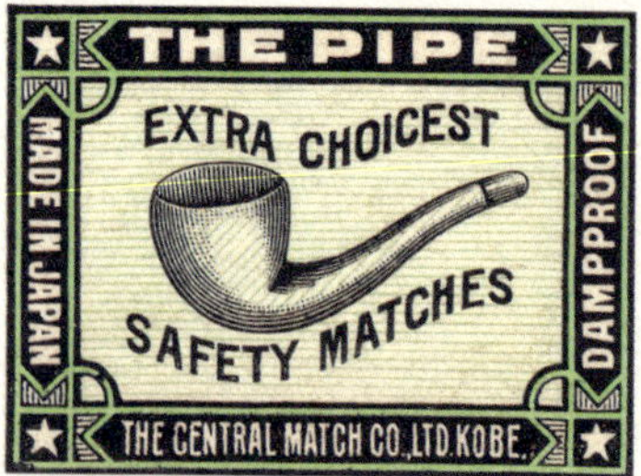
THE PIPE
EXTRA CHOICEST
SAFETY MATCHES
MADE IN JAPAN
DAMPPROOF
THE CENTRAL MATCH CO. LTD. KOBE.

BEST SAFETY MATCHES
烟斗商標
安全火柴
TRADE
MARK
CROSS PIPE

DOG AND CLOCK
頂上火柴
鐘狗老牌
天豊公司監製

The Alarm-Clock
Unequalled Solo
Safety Matches
MADE IN CZECHOSLOVAKIA

IMPREGNATED
GOBLIN
J.W.T.
J.W.T.
SAFETY MATCH
MADE IN SWEDEN

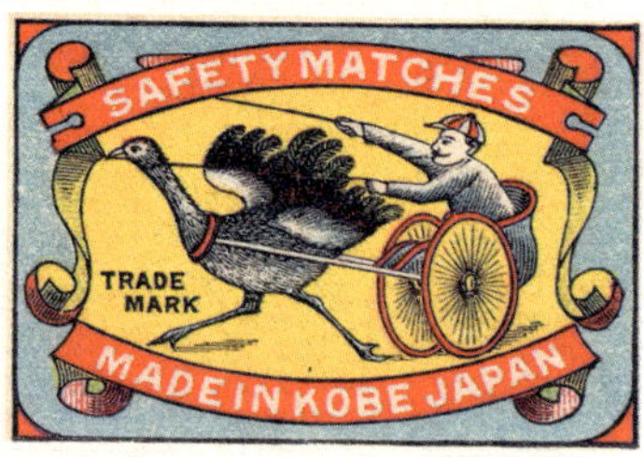
SAFETY MATCHES
TRADE
MARK
MADE IN KOBE JAPAN

STAR MATCHES
HIND MATCHES LTD.
SIVAKASI

मूनलाइट
SAFETY MATCHES
50'S-20Ps.

SAFETY MATCHES
FINEST
QUALITY
MADE IN SWEDEN

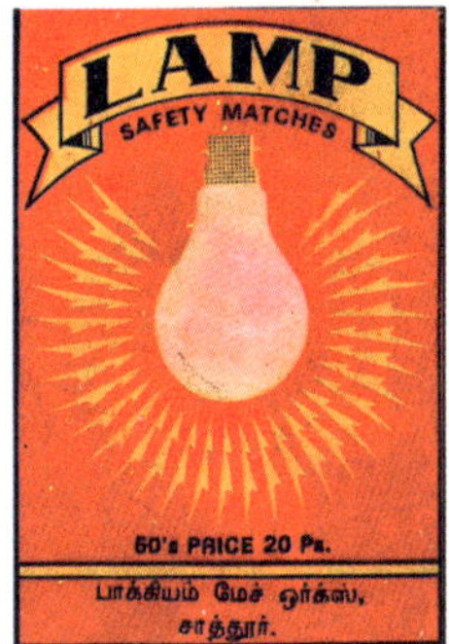
LAMP
SAFETY MATCHES
60's PRICE 20 Ps.
பாக்கியம் மேச் ஒர்க்ஸ்,
சாத்தூர்.

SUPERIOR
THE COCHIN MATCHES LTD, TRICHUR
SAFETY MATCHES

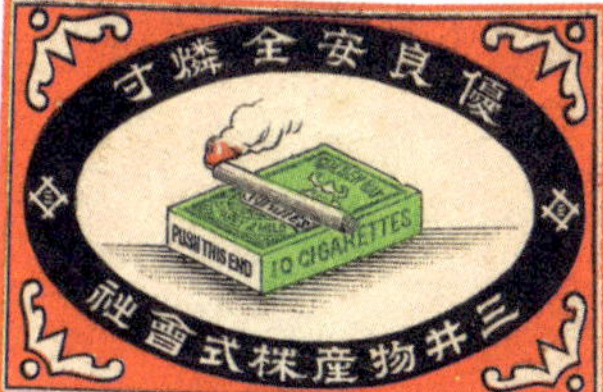
優良安全燐寸
PUSH THIS END
10 CIGARETTES
三井物産株式會社

TRADE MARK

RADIO
PRICE RE.0-0-6
SAFETY MATCHES
NATIONAL INDIAN
MATCH Co., DINDIGUL

THE PADLOCK
TRADE MARK
MADE IN ITALY
IMPREGNATED SAFETY MATCHES

アイロン印
優良適用燐寸

First published in the United Kingdom in 2026
by Skittledog, an imprint of Thames & Hudson Ltd,
6–24 Britannia Street, London WC1X 9JD

Designer: Alison Guile
Production: Felicity Awdry

EU Authorized Representative: Interart S.A.R.L.
19 rue Charles Auray, 93500 Pantin, Paris, France
productsafety@thameshudson.co.uk
www.interart.fr

A CIP catalogue record for this book is available from the British Library

ISBN 978-1-83776-119-7
01

Printed and bound in China by Starlite